Harvesting Coffee

The Life of a Coffee Bean

from Planting to Processing

By

Jessica Simms

The Journey of the Bean

Even if you usually go to a coffee shop for your daily cup, you could theoretically do most aspects of coffee preparation yourself at home, whether that's brewing, making your own blends, or even roasting your own green beans. The recent spike in interest in single-origin and other artesian coffees means this is even easier to achieve than it was a few years ago. Even relatively small towns often have at least one coffee shop that focuses on high-quality, single-origin beans, and you can purchase small batches of green beans—and the supplies you need to roast them—very easily online.

The growing and processing of the beans, however, is not something that can be undertaken by a hobbyist. Coffee plants are notoriously finicky; a slight change in the elevation, rainfall, or average temperature in the growing region can have a profound effect on the ultimate taste of the beverage in your cup. Most beans taste their best when grown at high elevations in tropical regions. Even if you could get a coffee tree to grow in a temperate region like Europe or the United States, the payoff would be a long time coming; while a coffee tree can produce beans for several years in healthy conditions, it doesn't usually start to produce its best-tasting beans until it's around 5 years old.

Even though you probably won't be able to start your own coffee nursery at home, learning about the journey a bean takes from the time it's planted until it ends up at the roaster can give you some excellent insights into why certain coffee tastes the way it does, and can help you to make more educated guesses about the flavor notes of a certain cultivar or growing region when you're buying unfamiliar beans, especially if you like to shop online and won't have anyone around that you can ask about the cup profile.

There are a lot of different factors that determine the quality level of the end cup. The genetic make-up of the plant will give it a certain starting quality and flavor potential, which will then be influenced by the environmental conditions during the flowering and fruit production stages of the bean's development, as well as by the farming practices in place and how the bean is processed and stored post-harvest. At each of these stages in the bean's life, the quality potential has to be carefully preserved to give you a good-tasting final cup.

The information that follows in this book will walk you through the life of a coffee bean, from the first moment the plant begins to grow until it's dried at the mill and ready to be shipped to the roaster. By the time you reach the end of this book, you'll have a much deeper understanding of exactly what has to happen to bring you your morning joe—and how farmers make sure the taste is exactly what you're looking for.

Historical and Botanical Background

In botanical terms, a coffee tree is any plant from the family Rubiaceae that produces coffee cherries (and, as a result, coffee beans). There are over 100 known species in this family, and new species are still being discovered on occasion, especially in the more densely forested regions of the Congo basin. Though all of these species are technically coffee plants, they are not all considered to be "true" coffees; that honor is reserved for plants in the genus Psilanthus or Coffea. This is largely a botanical differentiation, based mainly on the structure of the plant's flowers.

Even all the plants that are classified as true coffees are not necessarily suitable for grinding up and brewing into a cup. In fact, of the hundreds of species of coffee plant that exist, only two are regularly used for the preparation of the beverage: Coffea arabica (Arabica coffee) and Coffea canephora (Robusta coffee). Somewhere between 60% and 75% of the world's coffee production consists of Arabica, including the vast majority of coffees you find at your local coffee shop. If your favorite bean is a single origin, odds are good you're drinking an Arabica; even blends tend to be predominantly

Arabica, though some espresso blends incorporate Robusta since it can help to create a better crema. While you might hear different plant names being thrown around in regards to coffee like Bourbon or Geisha, these are referring to specific varieties or cultivars of the Arabica plant, not separate species; more information on those specific plants can be found in the next chapter.

Generally speaking, most people find Arabica coffee to have a better, more complex flavor, with a wider variety of potential flavor notes, a smoother texture, and a more pronounced aroma. In comparison, pure Robusta has a tendency to taste flat and doesn't offer the same sensory experience. The main advantage of Robusta over Arabica is its hardiness. It grows better in warmer climates and at lower elevations than Arabica, and is also more resistant to certain diseases that, most notably Coffee Leaf Rust (Hemileia vastatrix), a fungal infection that's a common affliction of plants throughout the coffee growing world. Hybrids of Arabica and Robusta can bring drinkers the best of both worlds, giving you both a complex flavor profile and an enhanced resistance to disease; the best-known hybrid is known as Timor (check out the next chapter for its typical flavor notes and cup profile).

The part of the coffee plant that is processed, roasted, and ground to prepare into a beverage is actually more accurately called a "seed" than a "bean." If you were to look at a coffee tree that's ready for harvesting, you wouldn't see anything that looks like a coffee bean; instead, you'd see clusters of yellow or red berries, known as coffee cherries. The bean used to make coffee is enclosed inside this fruit. There are a variety of ways this seed can be extracted from the fruit (for more information on that, check out chapter 7).

Coffee cultivation: a brief history

The evolutionary history of coffee is somewhat fragmentary. The coffee plant is indigenous to Africa, specifically the equatorial and highland forests of Ethiopia and South Sudan. Wild plants were first noted in the historical record around 850AD, and from there were spread to Yemen and the rest of the Arab world. Africa and the Middle East remained the primary domain of coffee until the early 17th century when European traders drank coffee in the ports of Istanbul and other major cities and decided to take this delicious concoction back to their home countries. A Dutch trader was the first to bring coffee back to Europe in the early 1600s. By 1650, the first coffee house was opened in Italy, and commercial plantations had been started in Indonesia and southeast Asia.

European explorers were also responsible for transporting the first coffee plants to the Americas. Legend has it that every coffee plant currently growing in the Americas is derived from a single clipping that was brought over on a ship from the Netherlands. Whether or not this is true, it is certainly the case that coffee did not exist in the new world until humans brought it over from Africa.

Even though it is not indigenous to the region, coffee plants quickly thrived in the mountains of South and Central America. Because importing beans from Colombia or Costa Rica to the United States is much cheaper and easier than bringing them all the way from Ethiopia, the North American market became dominated by beans grown in this region. Conversely, the European market was dominated by beans grown in southeast Asia, the Middle East, and Africa. This, more than anything, started the divergence between American and European (especially Italian) coffee drinking habits.

Today, coffee is grown in over 50 countries around the world, all of which fall within the inter-tropical belt. Although commercial coffee farms didn't start in the Americas until the 18th century, the land in countries like Brazil, Mexico, and Colombia proved so well-suited to growing coffee that the nations of this region soon dominated the international market, followed by plantations in Indonesia and south-east Asia, with the smallest percentage of the market share coming from Africa and the Middle East—ironically, considering this was where the plant originated in the first place. This is due more to the cultural traditions and farming practices at work in the various regions more than the viability of the climate or soil.

Cultivars and Varieties

All of the beans you can buy in a store or coffee shop come from the two major species of coffee (Coffea arabica and Coffea canephora). Within each species, there are smaller divisions, called varieties, each of which has its own ideal environment and flavor profile. A cultivar is a variety that has been purposefully produced through agricultural means in order to bring out certain characteristics from the parent plants. Most of the characteristics that affect coffee's quality—like bean size, acidity, flavor, and body—are highly heritable, meaning they're passed down consistently from one generation to the next. This means farmers can easily re-create a given cultivar once they've found one that has the qualities they're looking for.

Some cultivars occur in only certain regions of the world, while others are more widespread; some have very specific climate requirements, and others are known for their hardiness and adaptability. Because it's generally regarded to be a better-tasting coffee, the vast majority of cultivars and varieties that are sold as single-origin coffees are from the Arabica branch of the family. The notable exception to this is Timor, a hybrid of

Robusta and Arabica that's cross-compatible with either of its parent species. Timor is designed to have the taste of an Arabica along with the disease resistance and productivity of a Robusta and is especially popular in areas prone to coffee leaf rust. It may also be called Arabusta, especially when it's grown in Africa.

The pure Arabica varieties Typica and Bourbon are in many ways the backbone of the modern coffee landscape. Many of the most renowned cultivars in the coffee world are derived from one of these varieties. Both are derived from coffees that first originated in Yemen, although you'll find variants of the Typica and Bourbon everywhere in the world. Typica is known for its high quality, while Bourbon tends to be more productive.

The cultivar gives the coffee bean its potential flavor, which will then be altered and refined by the climate in which the tree is grown and the processing methods. The varieties listed below are some of the best-known in the coffee world. It is by no means a comprehensive list, but will at least give you an idea of the range of varieties that exist.

Heirloom

This is a designation that doesn't necessarily refer to simply one varietal but is worth including in this list because it does tell you something about where the plant came from. There are thousands of coffee varietals growing in Ethiopia. Coffee trees on most Ethiopian farms are able to cross-pollinate more easily than on monoculture farms, meaning that new varieties are always being created. This term has become a catch-all for any Ethiopian plant that hasn't been genetically tested to determine its specific variety.

Caturra

This variety is a mutation of Bourbon. It's named after the town in Brazil where it was originally grown and does best in a similar environment—a medium to high elevation with an annual rainfall of around 100 inches a year. While Bourbon is already a productive plant, Caturra produces even more beans of a more consistent quality, which makes it very popular despite the fact that it requires a lot of hands-on care and fertilization.

Catimor

This is one of the few varieties that has a bit of Robusta in its lineage. Catimor is a cross between Caturra and Timor and was developed to heighten the bean's disease and pest resistance. It has a darker flavor than Caturra, more bitter with fewer sweet notes, but it otherwise has a similar taste profile to Caturra and gives farmers a similar balance of production and quality even when it's grown at lower altitudes.

Mundo Novo

This is a cross between Bourbon and Typica that was first described in the 1940s. It was first grown in Brazil, and if you hear anyone referring to "Brazilian Bourbon coffee," this is probably the variety they mean. It has a relatively high disease resistance and grows best at a medium to low altitude. It has a slow maturation but a high production and can be planted in dense plots, which combines for a dense, flavorful bean that has a high yield. It is more closely related to Sumatra coffee than many American varietals, which gives these beans a more rounded overall flavor.

Geisha

You may remember the hype a few years back about this rare, wonderful coffee that cost about $80 a pound. That particular bean was a geisha, specifically one grown in Panama, and it is considered by some experts to be the best coffee in the world. One of the reasons it's so expensive is that the Geisha must be grown at high altitudes to reach its full potential, which means it has a relatively low yield and there aren't a lot of places that it can thrive. The taste is exquisite, with notes of citrus, jasmine, and honey and a heady, floral aroma, but it's the rarity as much as the quality that drives up the price. This varietal was originally a mutation of Heirloom and grown in Ethiopia. You can find geishas grown in a variety of countries; the ones from Panama are the most prized.

Blue Mountain

The Blue Mountain variety gets its name from the Blue Mountains in Jamaica where it is mainly grown. It's a variant of the Typica variety but with a higher disease resistance and better-suited for growth at high altitudes. The beans grown in Jamaica are in high demand since over 80% of the crops are exported to Japan, making it one of the world's most expensive varieties; the same variety grown elsewhere in the world gives a similar cup flavor but at a significantly lower price. The mild flavor, low acidity, and low bitterness are what make this coffee so admired in terms of taste. When it's grown outside of Jamaica, it may not bear the Blue Mountain name; look for PNG Gold out of Papua New Guinea or Boyo out of Cameroon.

Maragogype

This is a mutation of the Typica variety that was discovered in the Bahia region of Brazil. Maragogype beans are significantly larger, giving them the nickname "Elephant Coffee Beans." The quality of Maragogype depends heavily on the soil in which it was grown. Good Maragogype has a mild flavor with a clean finish and a low acidity; lower-quality beans can have a flat taste and poor aroma.

French Mission

This is a name given to a variety of Bourbon that was planted by French missionaries in Africa, giving it its name. It was first planted in the late 19th century and likely has some Mocha qualities as well from natural mutations that occurred growing alongside beans brought from Yemen. Many of the most prized Kenyan varieties were derived from this bean, notably K7, a version of French Mission grown primarily at the Legetet Estate. French Mission beans tend to have a flavor that's sweet and nutty, with a mild overall cup profile and a smooth texture.

Ruiru 11

Another Kenyan coffee is the Ruiru 11 cultivar, which was developed at Ruiru station in the 1970s. The initial intent was to create a highly resistant strain of coffee by crossing Timor (resistant to leaf rust) and Rume Sudan, a variety with resistance to coffee berry disease. The bit of Robusta added to the mix from the Timor also gives this coffee a high yield, but there is some controversy still about the taste effect on the final cup. Some argue that it is a naturally less complex cup because of its genetic lineage; others insist that inconsistencies in the taste of the bean are caused by a lack of pruning from the farmers, who want to get the highest possible yield out of their crop. The generally held opinion in the professional coffee world is that Ruiru 11 has some of the dry berry notes Bourbon is known for but gives a less flavorful overall cup than other Kenyan varieties.

SL28 and SL34

On the other side of the Kenyan equation are the cultivars SL28 and SL34. Both of these were developed by Scott Labs in the early 20th century. You can identify beans developed by Scott Labs because their name will start with the designation "SL" followed by the number. It's believed up to 90% of beans currently grown in Kenya were derived from a Scott Labs-developed bean. The 28 and 34 varieties, in particular, have gained quite the reputation among coffee aficionados. SL34 was based on French Mission and has a generally higher yield and can grow at lower altitudes than many Scott Labs cultivars. SL28, meanwhile, is admired for its combination of drought resistance and high cup quality, with a black current-like acidity and a clean overall flavor profile.

Environmental Conditions

As was mentioned in chapter 1, both Arabica and Robusta originate from Africa—specifically the equatorial rainforests in the northern half of the continent. The exact conditions that are ideal for each of these species are different, however. Arabica typically grows best with an average temperature of between 64°F and 70°F, at an elevation of 3,900-6,400 feet and with an annual rainfall of 43-79 inches. It also has specific soil conditions that are most conducive to its growth—typically deep soils that are red or reddish-brown and have good drainage, with a pH of 4.1 to 6.3. In terms of where you'll find these conditions, the ideal agro-ecological zone for Arabica is between around 20° north latitude to around 25° south latitude.

The ideal growing conditions for Robusta are a bit different. Like Arabica, it prefers reddish soils that are well-drained and either flat or have a very gentle slope. Robusta grows best in fairly acidic soil that has a low native fertility. In terms of climate, Robusta likes an average temperature of around 71°F-79°F with an annual rainfall of 47-98 inches, and it prefers a lower elevation than Arabica, somewhere in the neighborhood of 800-4,900 feet. The ideal agro-ecological zone is slightly reduced compared to Arabica, spanning around 15° north to 15° south latitude. As you can see, Arabica is better adapted for cold, while Robusta is better adapted for the heat.

This environmental preference of each of the plants is part of the reason they tend to differ in terms of taste and complexity. In general, the higher the elevation at which the plant is grown, the cooler the conditions and the more slowly the fruit and enclosed bean will mature. A slower maturation process results in a denser bean with a more intense flavor. This has to do with which volatile compounds like acetone and ethanol are allowed to develop within the bean; these are the compounds most responsible for qualities like aroma, acidity, and fruity flavor notes.

Of course, there are not only exceptions to these general ranges of temperature, rainfall, and elevation, but there are also significant variations within the range that can have a major impact on the overall flavor of the cup. Various regions of the coffee growing world have different combinations of soil quality, elevation, rainfall, and temperature. These can generally be grouped into five categories: hot-dry, hot-wet, constant, cool-variable, and cool-dry.

Hot-dry

Around a quarter of the world's commercial coffee comes from a hot, dry region of the world. These regions tend to have a relatively low elevation, around 2,500-3,000 feet, making it a generally better environment for Robusta than Arabica, though there are exceptions to this rule. Hot, dry climates are found in western Africa and regions of Brazil. Beans tend to have a relatively low density and require heavy use of shade to bring out the best flavors.

Hot-wet

Hot and wet climates tend to be found in low to medium elevations, and though less of the world's coffee is grown in these environments than in hot, dry climates, they are found in every major region. As with hot, dry climates, it tends to be a better environment for Robusta and Arabica. It can be found in Central African countries, certain regions of southeast Asia (especially Vietnam), and coastal and island regions throughout Central America. Coffee in these areas tends to be very fast-growing, giving the farms high yields, though the flavor is more subtle and they tend to have less aroma.

Constant

You could say that a constant environment provides the best overall climate for your basic, very good cup of coffee. The beans grown in these environments will not be spectacular, like the coffees grown at very high altitudes, but the trees will strike an excellent balance if yield and flavor. About a quarter of the world's commercial coffee crops are grown in constant environments, the vast majority of which is Arabica.

Constant climates are found in areas like Costa Rica, Colombia, Ethiopia, and Indonesia, where the relatively stable equatorial weather is balanced by the medium to high elevation of the farm. A wide range of different flavors can be brought out of the beans grown in this environment, largely through different farming and processing methods.

Cool-wet

A fifth of the world's coffee comes from a cool, wet environment—including some of the most highly-prized coffees in the world. Cool, wet environments tend to be found at high elevations, and the beans grown in them are dense with a complex flavor, texture, and aroma. Many African nations have a cool, wet environment, including Uganda and Kenya, as do South American nations like Peru and Colombia.

Cool-dry

Also tending to come from higher elevations, the cool, dry environment is the least common among commercial coffee crops, probably in part because coffee grown in this environment has the lowest per-tree yield of any of the different climate zones. Like cool, wet environments, the beans grown here will be very dense and have an intense flavor.

Climate stressors

Even when it's grown in a region known for producing great beans, mother nature doesn't always cooperate by providing the right climate. Coffee farmers have to be able to adapt to weather conditions when they're not ideal. Climate-related stress can be responsible for reduced growth or damage to the leaves, issues with the tree's metabolism, or in extreme cases even tree death.

Excessive high wind can cause damage to the branches and leaves or even strip the leaves completely off the plant. Planting trees around the coffee plants can help to provide windbreaks that will reduce the potential of damage to the coffee trees. Pruning the coffee plants regularly also helps to strengthen the branches, making them less susceptible to damage.

Drought is another common affliction of coffee plantations. Even when it doesn't lead to crop failure, drought can reduce the size of both the trees and the fruit they produce, resulting in a lower yield. Irrigation systems can help prevent both drought and flooding by regulating the flow of water. Temperatures that fall outside the normal range can also impact crops. A higher plant density can be helpful in regions that sometimes experience low temperatures, while regions that can get too hot will benefit from the use of shade trees, which can help regulate the temperature of the coffee trees.

Long-term changes in the climate of a region also have a big impact on the production of coffee trees. In recent years, there has been a trend of elevated temperatures in many of the world's coffee growing regions, especially in the South and Central American nations. This is causing some land previously suitable for coffee growing to become unusable, and is likely a contributing factor to the increased severity of infections like coffee leaf rust in the region. It's uncertain what additional challenges this will present to farmers in the coming years.

Farming Methods

The farming methods in common practice in various areas of the world are determined by a variety of factors, including local cultural traditions, climate conditions, and the topography and condition of the land. Regardless of the specifics of a certain farm's practices, the end goal is the same: to maximize the yield and quality of the beans while minimizing labor and production costs. This can lead to a very interesting pastiche of traditional and modern farming practices in many areas of the world.

Some factors that influence the taste of the final cup are outside of the farmer's control, like the elevation and topography of the land. In most cases, though, there are ways that the farmer can compensate for even unpredictable aspects of the plant's growth, like rainfall or temperature, through the careful management of resources and use of modern farming technology. While each farm's practices will be unique, there are four major aspects that coffee farmers focus on to improve their crops: plant selection, fertilization, shade, and irrigation.

Plant selection

The general category of plant that a farmer will use is by and large determined by the conditions. Certain varieties and cultivars are better suited than others to certain elevations and climates; part of a farmer's job is to choose the plants that will be best able to thrive on his land. Along with choosing which cultivar he wants to plant, a farmer has to select the right propagation system for trees he plants on his land in the future.

There are two general ways that a coffee plant can be reproduced: generative and vegetative. Generative refers to the growth of a plant from a seed and is the easiest and most natural propagation method. It will produce trees with a stronger root structure, but also will take longer for the tree to reach maturity, and is less predictable, with more possible variation from what's expected of the variety.

Vegetative propagation, on the other hand, involves using portions of an already-grown coffee tree to start a new plant, whether that's achieved by planting a cutting or by grafting a branch onto an existing tree. Cuttings are especially useful for large coffee plantations that are hoping to achieve a uniform taste and quality out of an entire crop. They are the best way to produce a new plant that has almost identical taste characteristics to those already on the farm.

Grafting, on the other hand, is more often used to create hybrids between cultivars or varieties. One common practice is to graft Robusta branches onto an Arabica tree to increase the yield or disease resistance without sacrificing the complexity or aroma of the cup. Generally speaking, grafted plants are also hardier and better able to withstand extreme climate conditions like droughts or heat waves. On the negative side of things, they also run the risk of having compatibility issues, which can ultimately cause the grafted branch to fall off of its new tree or even lead to the degradation of the plant as a whole.

Fertilization

One of the key factors in maintaining a successful farm of any kind is to preserve the fertility of the soil. While other crops benefit from rotation, the long lifespan of a coffee tree makes this impractical. Instead, it is up to the farmer to replenish the nutrient level in the soil so that the plants which grow on it will continue to be healthy in the long term.

Several factors determine the fertility of the soil. Some of these are physical, like the texture and depth of the effective soil, or its level of aeration or water infiltration. Others are chemical, like pH, salinity, toxicity, or the ratio of nutrients. Finally, the other creatures that make use of the area will impact the soil. Leaves dropped from other plants, droppings from animals, and microorganisms living in the ground will leave their traces in the soil, which will make their way into the coffee trees.

The most important nutrients for the production of coffee are nitrogen, phosphorus, and potassium, and to a lesser extent calcium, sulfur, and magnesium. Using zinc in the fertilization mix helps to improve bean size consistency and can reduce the instance of coffee berry borer infestation. Finding ways to replenish these nutrients in the soil is important to maintaining the quality of the land and by extension the plants grown on it. Many farmers will regularly sample the soil and analyze its composition to make sure it has the right nutrient balance. This is one-way modern technology can be a great benefit, with both hardware and software available on the market that can make soil analysis easier and more precise.

It's also very important to maintain the right level of acidity in the soil. This is one trait that can have a significant impact on the cup's flavor profile, especially important for bringing out fruity notes. Fertilizers will often have some effect on the pH level of the soil, but there are other compounds that can be used to regulate it, like dolomite, gypsum, and lime.

Shade

Because it evolved in the undergrowth of tropical forests, the coffee tree is naturally shade tolerant. Too much direct exposure to sunlight can bleach nutrients out of the leaves and reduce the complexity of the finished cup, making it bland or even giving it notes of carbon and leather. The traditional arrangement of a coffee farm mimics the environment in a tropical forest, with trees of other species planted between the rows of coffee plants. These not only helped to fertilize the soil by dropping leaves and other organic matter but also provided natural shade.

Starting in the 1970s, large commercial farms started using what was called a full-sun monoculture system, in which the entire plot of land was devoted entirely to coffee trees and the area around them deforested. While this helped to reduce the instance of some insect and fungal pests—especially coffee leaf rust—it also left the coffee plants constantly exposed to direct sunlight.

In the modern era, most farmers find the best tactic to be somewhere in the middle, interspersing designated shade trees in between the rows of coffee plants. A good rule of thumb is around 40% shade cover; any more, and the coffee trees won't get enough heat, reducing the yield of the plants. A mix of species native to the region can also help to keep the soil in its natural balance. By strategically employing shade in this way, farmers can slow down the tree's metabolism, lengthening the ripening process so the flavor in the bean has more time to develop.

Irrigation

Many of the nutrients that end up in the coffee bean are brought in with the water the roots absorb from the soil. Making sure the soil has the right saturation level is important to controlling the rate of the plant's maturation and the ratios of compounds within the bean. Soil that's too wet can also make the trees less stable, loosening the hold of the roots, and can contribute to the proliferation of insects or fungus that can weaken the tree. On the other side, if the soil's too dry the tree will produce smaller fruit and fewer leaves, ultimately resulting in a lowered yield and quality.

Irrigation systems are the easiest way to make sure the water level in the soil stays consistent throughout the growing season. The topography of the region will impact the exact system and arrangement that's used. In cases where the elevation is steep, a terraced system might be the best solution; on flatter lands, systems that run up and down in rows are more common.

Not all farms will need much in the way of irrigation. In wetter climates, the natural rainfall levels may be sufficient to supply the soil with enough water; in these areas, it's ensuring proper draining that's the main concern to farmers. Planting native trees with deep root systems can help to hold the soil in place and keep it from flooding even during heavy rain.

Sustainability

There are some regions in which sustainable and organic farming are almost a given, not because of any regulations of the local government but because farmers in the area either can't afford or don't have access to chemical pesticides and other modern techniques. This is especially true of many African nations, where the long history of coffee cultivation also encourages the use of traditional methods for cultural regions.

Other coffee-growing areas of the world favored a monoculture approach for much of the twentieth century. Farmers would clear the native vegetation completely from the land to make way for the coffee trees, then use an irrigation system to see to soil drainage and chemical fertilizers to return nutrients to the soil. While these techniques can effectively increase a farm's production, they ultimately have a negative impact on the flavor of the coffee. Coffee is one of the few crops that not only can be grown in harmony with its environment but in fact, thrives under a natural rainforest canopy. The trees provide shade for the leaves, slowing the maturation and allowing the beans more time to develop. The roots of large trees also help with soil drainage, while the biomatter dropped by the surrounding plants replenish the nitrogen and other nutrients absorbed by the coffee trees, eliminating the need for costly commercial fertilizers. While many large plantations in both Indonesia and South America continue to use a monoculture approach, there has been a definite shift in recent years toward a more sustainable approach by smaller farms.

The use of chemical pesticides is a trickier question. Part of the reason the sustainable growing movement has been met with such enthusiasm by coffee farms is because it provides a tangible benefit to their business. Beans grown in a natural environment are generally higher quality, and while the per-acre production may drop, the farmer can also save money on fertilizers and irrigation systems. Commercial pesticides, though costly, are also markedly more effective at preventing devastating insect infestations than any natural method. Considering that some of these infestations can easily wipe out an entire farm, it's understandable that many farmers are reluctant to give up the one proven method of preventing this destruction.

It is yet to be seen how the changing climate in the future will affect these efforts toward more sustainable and organic farming practices. The use of hybrids like Timor or grafts of Robusta onto Arabica plants can improve the resistance of the plants to diseases like coffee leaf rust, and can also help the plants to adapt to warmer temperatures. While coffee purists might object to these practices on a taste basis, it may become more and more common in the coming years as farmers adapt to their changing landscape.

Coffee Growing Regions

There are three general coffee growing regions in the world: Africa, Central and South America, and Southeast Asia. These are broad areas, obviously, between them encompassing over 50 different countries, so you can't necessarily predict how a coffee will taste just because of what continent it comes from. Despite the variations, there are some general traits you can expect from a given region's coffee, mostly due to what varieties, farming technology, and processing methods are available to the farmers.

The sections that follow will give you an overview of what defines the coffees from the world's various regions. Each section starts with a general description of that regions overarching traits, followed by a more detailed description of the coffees from the most popular or noteworthy countries. As you'll see, some regions and countries show more variation than others, but all have their own distinctive traits and profiles.

South and Central America

Three of the five largest coffee producing countries in terms of pounds sent for export are in this region of the world. Especially for the North American market, it is the region that best exemplifies the "standard" cup of good coffee. Good Latin American coffee is known for having a clean finish and a bright acidity, but there is a lot of variation in cup character depending on the exact nation of origin. High-grown coffees from both Central America and northern South America tend to be bold with a full body. Lower-grown American coffees have a softer, rounder flavor, and can be sweet or dark.

While there are naturally-occurring mutations that have come out of South and Central America—many of which are now popular variations in their own right—the planting of coffee trees on farms in the Americas is very controlled, with the farmer choosing the right cultivar to suit his soil, climate, and elevation. There is also a lot of control on the processing side of things. Though most coffee farms in this region are small, family-owned operations, the majority of Central and South American coffees are wet-processed in centralized mills, which helps to give them their clean and consistent finish.

Brazil

Brazil is the world's largest coffee producer, something that is at least in part due to the extensive geographic area suitable for growing coffee in the country. Brazil is also the most atypical American coffee-growing country in terms of the farming and production of the beans. Coffee farms in Brazil tend to be large plantations which employ hundreds of workers, often with their own on-site processing facilities, which leads to more variation in the bean's flavor. It is also not uncommon for these farms to employ multiple production methods, sometimes within the same harvest, giving the beans even more variation.

Brazil's large area also means a lot of variation in climate throughout the country. Both Arabica and Robusta are grown here, with the farm's elevation largely determining which species of coffee will grow the best in a given area. This diversity of varieties allows for more naturally-occurring mutations in Brazilian coffee farms than elsewhere in the region; many varieties that are now popular elsewhere in the world were first discovered in Brazil.

In general, growing areas in Brazil are at a lower elevation than the rest of the region (around 2,000 to 4,000 feet) which gives them a relatively low acidity and mild, more nuanced overall flavor. The typical profile of a good Brazilian coffee is a sweet, medium-bodied cup with a clear, clean finish. Coffee from the Bahia region is known for its impressively large beans; other popular growing regions are Mogiana, Sul Minas, and Cerrado.

Colombia

The world's second-largest coffee producer, Colombia is arguably the world's best-known coffee-growing region, especially from the perspective of the North American market. Colombia exemplifies American coffee both in terms of process and taste. Beans are grown on small family-owned farms, and there are thousands of these farms throughout the country. The landscape tends to be at a high elevation and very rugged—the perfect environment for coffee growth, though less so for transportation, and even to this day beans are often transported by jeep or even mule out of necessity.

The highest grade of Colombian coffee is called Colombian Supremo. It has a delicate flavor with sweet notes and a high aroma. The second-highest grade, Colombian Excelso, has a similar flavor but with more acidity. Good Colombian coffees are known for their balance more than any specific flavor notes.

Peru

Like Colombian coffee, Peru's coffee farms tend to be small, family-owned, and at relatively high elevations. Most Peruvian coffees are wet-processed, with the best beans coming from the Chanchamayo and Urubamba Valleys. The country employs its own unique grading system for beans; most beans you'll find in specialty shops is of the highest grade, AAA.

You'll often hear Peruvian coffee described as being good for blending because of its understated character. The best beans have a light, mild flavor with sweet, nutty notes, a generally low acidity and light body, and tons of aroma. Lower-quality Peruvian beans are often used as dark roasts or as a base for flavored coffee.

Costa Rica

The largest coffee producer in Central America, Costa Rican coffee is perhaps the most consistent of any nation's. The topography of the region is mountainous, with a high elevation and relatively wet climate. All of Costa Rica's exported beans are wet-processed Arabicas. These are grown on small farms then immediately sent to centralized processing facilities, known as beneficios. While there is some variation in the drying process (you'll find both mechanically- and sun-dried offerings) the rest of the process is relatively consistent across the country. You'll find it has a sharp, bright acidity and a medium body, and the overall cup tends to be well-balanced with notes of chocolate and citrus.

Guatemala

Like Costa Rica, Guatemala has a rugged landscape that provides volcanic soil at high elevations of 4,500 feet or more.

This results in a bean that's generally medium or full-bodied with spicy or chocolatey notes and a complex, lingering flavor. Because there are many microclimates in the country, the specifics of the cup will vary region to region. The three main growing regions are Antigua, Coban, and Huehuetenango; you'll see beans from all three on the North American market.

Nicaragua

The late 20th century was a difficult time for Nicaraguan coffee farmers. Not only did the country go through a brutal civil war, Hurricane Mitch ravaged the nation's farmland in the late 1990s. It is only in recent years that Nicaraguan coffees have become widely available again on the export market. The best beans come out of the Jinotega, Matagalpa, and Segovia regions. They're typically grown under shade at high altitudes and have a fragrant complexity, with notes of both nut and chocolate.

Mexico

Though its output is lower than countries like Colombia and Brazil, Mexico still produces more coffee annually than any of the African nations. Despite the high volume they send for export, small farms are still more common than large plantations. Most of the farms are in the southern states, including Veracruz, Oaxaca, and Chiapas, where the topography is very similar to the mountainous terrain of Central American coffee-growing countries.

Coffees from Mexico are generally deeper in flavor than South American coffees, with a sharp finish and a pronounced aroma. This depth of flavor makes them especially well-suited for dark roasts and blends since the taste of the coffee is strong enough to shine through the roast level or added flavors.

Hawaii

The only state of the United States where coffee can grow is Hawaii. There are farms throughout the Hawaiian Islands but the best-known ones are on the big island of Kona, where coffee is grown on the slopes of the Mauna Loa volcano. This black volcanic soil gives these beans a high acidity level, while the tropical clouds provide afternoon shade and frequent rain showers, a combination of environmental conditions that give Hawaiian coffee a rich aroma and body.

Africa

Compared to the control exercised at all stages of the process in Central and South America, coffee from Africa is a much looser when it comes to harvesting and processing methods. You'll find a lot more variation from country to country in terms of how the beans are grown, picked, and dried, from the large estates of Kenya to the wild coffee trees of Ethiopia.

Despite this variation, there is a flavor profile distinctive to coffees from Africa and the Arabian Peninsula. These coffees tend to have very assertive fruit, floral, or wine notes, often emphasized in African coffees that have been natural processed. It is believed to be caused by the composition of the soil, though even coffee experts still aren't completely sure just what it is about the African soil that imparts these notes.

Ethiopia

The birthplace of coffee, Ethiopia is the only country where wild coffee forests remain a prominent harvesting source. It is the largest producer by volume in Africa and the sixth largest worldwide. Ethiopian coffees are among the most distinctive and varied, with a lot of the flavor differences depending on how the picked cherries were processed. Most Ethiopian coffee is grown on small family farms, under shade in interplanted lots and without the use of any chemicals. The exception to this are the larger government-run estates in the southwest of the country.

There are three main regions in Ethiopia: Harar, Sidamo, and Yrgacheffe. Both Sidamo and Yrgacheffe coffees that make it to export are wet-processed, prepared at central washing stations in a similar system to that employed in Costa Rica. The only beans that tend to be dry-processed in this region are small batches intended for local consumption, which tend to be less consistent and a lower overall quality.

The Harar region is slightly different. Beans of all qualities are typically prepared in a natural dry process, giving these beans a slightly fermented aftertaste and a complexity that many aficionados find appealing. Regardless of region, good Ethiopian beans tend to have a full body with assertive fruity or winey notes.

Kenya

Coffees from Kenya are the most widely popular of any of the African varieties. They're known for their sharp acidity, full body, and rich aroma. The country employs its own 10-size grading system, with AA being the largest, and in general maintains tight control over the processing and drying methods, which use the most advanced techniques and are generally very technically sophisticated and export crops are sold in an auction system. The overall result of all these factors is a consistently high-quality product.

Kenyan beans are also consistently high-grown. There are two primary growing regions in the country: on the foothills of Mount Kenya just outside the capital city of Nairobi, and on the slopes of Mount Elgon further south in the country. Most of the beans are grown on small cooperatives, though they may share a centralized processing facility.

Yemen

Though Ethiopia was the first place coffee was grown, Yemen was the country to cultivate it commercially. Farms today use many of the same techniques that were employed hundreds of years ago, with trees grown in small terraced gardens. Because water is relatively scarce, the beans tend to be smaller and more irregular, and are always dry-processed. The resulting taste is deep and rich, which earned the beans the moniker "Mocha," which is most commonly seen today when the bean is used in Mocha-Java blends.

Tanzania

The majority of Tanzanian coffees sold in the United States are peaberries, the genetic mutation that merges the two halves of the coffee seed into one dense bean. Most of the country's beans are grown on the slopes of Mount Kilimanjaro or Mount Meru, near the border with Kenya. The beans have a soft and floral flavor, with a rich acidity. It's graded similar to Kenya, with AA being the highest grade and B the lowest.

Burundi

Burundi is a relative newcomer to the North American coffee market. It uses a similar auction-style export process to Kenya, which means the beans that come out of Burundi are typically of a high quality and sell at a relatively high price point. Most of the beans coming out of Burundi are shade-grown and organic, giving them a bright acidity with floral notes and aroma. They're usually grown on small farms in the north of the country then sent to centralized washing stations for processing.

Uganda

The Robusta variety was originally discovered in Uganda, and this remains the bulk of the country's coffee production. The Robusta grown here is typically dry-processed and is rarely sold as a single-origin variety, instead used by commercial coffee manufacturers as filler in blends. While Uganda remains best-known as the birthplace of Robusta, you can also find a few wet-processed Arabicas from the country. The most admired of these is the Bugisu variety, which is grown on the slopes of Mount Elgon near the Kenyan border and has a similar taste, with wine and fruit notes, though it's a bit rougher in texture.

Southeast Asia

The best-known coffees from this area come from the Malay Archipelago, which is the chain of islands that contains the countries of Indonesia and Papua New Guinea. There is perhaps the most variation in terms of both tastes and processing methods in this region, which can have a bright fruitiness or a deep earthiness depending on the farm it was grown on.

Coffee farms in Southeast Asia take one of two forms: either large estates, which often use the most advanced technology for growing, harvesting, and processing the coffee, or small holdings, which are often more traditional operations. Coffees from smaller farmers will typically have more earthiness and may have a slightly fermented flavor; those from larger estates tend to produce a cleaner cup.

Indonesia

Indonesia is third in total coffee production behind Brazil and Colombia. The country made up of many different islands in the Malay Archipelago, three of which produce the majority of the nation's coffee: Java, Sumatra, and Sulawesi. The beans that come from these three regions is distinct enough that you'll typically see it labeled for its island of origin rather than the country as a whole.

There was a time that Java led the world in coffee production until practically its entire crop was wiped out by coffee leaf rust in the mid-twentieth century. The majority of the land was then re-planted with Robusta because of its high resistance to the disease. More recently, Arabica has made a reappearance on the island, mostly on the eastern side of it. Most of this Arabica is grown on large farms that use the most modern equipment for processing and drying, giving the beans a cleaner profile than other Indonesian coffees, with a taste that's bright, sweet, and fragrant.

Coffee from Sumatra, on the other hand, tends to come from small plots where the trees are grown without shade or pesticides. The processing method is in theory very similar to that used on Java—traditionally a washed or semi-washed coffee—but in practice, the Sumatran process is less consistent, with the beans frequently allowed to ferment for longer before being pulped. This results in a coffee with a very intense and complex flavor, rich in body without being heavy. The best-known regions are in the north of the island and include Linton and Mandheling.

Sulawesi may also be called Celebes or could go by the name Toraja (the indigenous people of the island) or Kalossi (the market town most bean harvests pass through). Most of these beans are grown in the mountains near the port of Ujung Padang. Of the Indonesian coffees, Sulawesi has the most variation. The beans grown on small farms are typically dry-processed and have an earthy, pungent flavor reminiscent of a Sumatra. Those grown on plantations are wet-processed and result in a vibrant, smooth cup more similar to Java. Sulawesi is also known for its aged coffees, which are allowed to sit for a time in the warm, damp climate, a process that lowers the acidity and gives the coffee more body.

Papua New Guinea

Like other beans from the region, coffee grown in Papua New Guinea comes in two distinctive versions: estate-grown coffees that are processed in large facilities and small holder coffees that are processed on-site by the farmers. Both versions tend to be grown organically in similar climates and elevations and are wet-processed, but the estate-grown coffees are generally more consistent, with a vibrant clarity that makes the beans well-suited to a light or medium roast. Variations in the drying and shipping methods can give small holding coffees an earthiness or fermented aftertaste.

Vietnam

You're unlikely to see any coffees from Vietnam on the shelf at your local specialty coffee shop, yet the country is the fourth largest exporter of coffee by volume in the world. The reason you likely haven't heard of Vietnamese coffee is because the majority of the beans grown in the country are Robusta, not Arabica. Vietnam's Robusta tends to have a lower acidity, milder body, and better balance than that grown in other regions of the world. This makes it especially popular for use in blends, especially for use in espresso machines, where the addition of Robusta can help to improve the crema.

Pest Control

One of the primary struggles of any farmer is to prevent insects, fungus, and other life-forms from interfering with their crops. This is as true of a coffee crop as any other, and there are some coffee ailments that can be especially devastating to commercial crops. The leaf rust outbreak of 2013 stretched from Brazil all the way to southern Mexico and caused many small farms in Central America to go out of business. Affected regions reported crop losses of up to 70%, and some regions of Guatemala, Honduras, and El Salvador were forced to declare states of emergency due to the sheer percentage of the population that found themselves out of work as a result. The international coffee market is still seeing elevated prices some four years later as a result of this infection.

There are 16 recognized fungal diseases and 19 recognized insect pests that affect coffee plantations. Because of the prevalence of these various parasites, the majority of coffee farmers use some form of pesticide on their crops to protect their livelihood. A certain amount of pesticide use is permitted even in coffees grown certified Rainforest Alliance or Fair Trade Organic, provided safe application practices are used and the products in question are demonstrated as non-harmful to the surrounding environment. While many people have an automatic aversion to the idea of pesticides, the truth is without them the majority of coffee farms around the world would be unable to exist. Because the coffee bean is enclosed in the fruit when pesticides are applied—and then goes through a lengthy preparation process before even reaching the roaster— the likelihood of any chemical traces making it into your finished cup is practically zero.

Fungal and insect parasitism during the maturation process of the bean is the number one cause of defective beans. Coffee tasters with especially well-trained palates can often taste the specific type of ailment that afflicted the plant when they encounter such defective beans in their brewed cup. Check out the rest of the chapter below for more specific information on the most common coffee parasites and what impact they have on the bean's development and taste.

Insect pests

There are two main ways insects cause damage to a coffee plant: by consuming it or by laying its eggs within the plant. While both of these activities are more likely to be aimed at the leaves and fruits than at the seeds themselves, these attacks can reduce the overall health of the plant, making it more susceptible to other infections and reducing the types and

43

amount of nutrient compounds that are stored within the bean, thereby affecting the flavor.

The main insect pest that does damage the beans themselves is the coffee berry borer (Hypothenemus hampei). The coffee plant is the only host for this species, and it can detect coffee plants from a long distance away. When it finds a host plant, the borer tunnels through the cherry into the bean, where it lays its eggs. The larva damages the structure of the bean as they're eating their way out. The tunnel bored by the adult to lay the eggs also frequently allows the growth of fungus and other microorganism infections. Beans affected by the berry borer have a reduced density and tend to have a fermented or moldy taste in the final cup.

Thanks in large part to the use of pesticides, the infestation level of the coffee berry borer is currently less than 5% worldwide. The use of hybrids and grafts can also increase the resistance of coffee plants to the borer, and these practices are widely utilized in high-risk areas.

While the coffee berry borer is the most pervasive insect pest to affect coffee plantations, it's certainly not the only culprit. Several species of insect are known to feed on or lay their eggs on coffee plants, including Antestia bugs, mealy bugs, and several species of fruit flies. There are also insects that can affect the beans after harvesting. The coffee weevil is the best-known of these; it feeds on beans when they're in storage, making them unsuitable for roasting if the infestation isn't caught and eliminated early enough.

Fungal parasites

Of the 16 recognized fungal coffee diseases, five directly attack the cherries: coffee berry disease, oily spot, pink disease, berry blotch (also called iron spot), and American leaf spot, frequently known by the Spanish name Ojo de Gallo. Berry blotch is the most common of these and contributes to malnourishment in the coffee plant, especially nitrogen deficiencies. Berry blotch is most common during dry periods. It causes an early ripening of the cherry, leading to red lesions on the skin that eventually turn brown and necrotic by the time the fruit matures. It can be reduced by increasing the shade the coffee plant receives; this slows down the plant's metabolism, giving the fungus less opportunity to thrive.

Some fungal infections are specific to a growing region. Coffee berry disease afflicts only African crops. It invades the cherry during the first and second stages of fruit growth, around 4-14 weeks after the plant flowers. The beans themselves are unaffected by mild attacks, which leave a scab tissue (also called "cork tissue") on the surface of the cherry and largely affect only the pulpy tissue of the fruit. A more severe infection is indicated by dark brown spots which will eventually cover the entire surface of the cherry. An infection of this magnitude will affect the development and quality of the bean.

Conversely, American leaf spot is an ailment exclusive to the coffee plants of South and Central America. It tends to occur

45

when the temperature and humidity are consistently higher than average for the region. Its effects can be prevented or mitigated by reducing direct exposure of the plant to sunlight with careful use of shade.

The most damaging fungal infection that can afflict a coffee plant doesn't attack the cherries directly, but rather the leaves. Known as Coffee leaf rust, this infection causes extreme defoliation and shedding, and can eventually lead to the death of the tree itself if left unchecked. If the infection occurs in the first two stages of cherry development, it can cause them to fall from the tree. In later stages, the cherries themselves may be unaffected, but the flavor of the bean will still suffer due to the leeching of minerals and nutrients through the damaged leaves.

A fungal infection of the cherry plant can cause a distinctive off flavor in the finished cup. The infection causes the sugars inside the beans to degrade, often replacing them with fungal metabolites that can give the cup a woody, pulpy flavor or a harsh, astringent feel in the mouth. While fungal growth stops once the humidity of the bean drops below 12%--meaning it's halted by the drying process—any damage done before this point cannot be reversed and will continue to affect the flavor of the final cup.

Plucking and Sorting

With crops that grow in temperate regions, the seasons dictate the planting and harvesting times. With a crop like coffee, though, that grows in a more tropical environment, the exact number of times a tree is picked over the course of the year and which months are included in the growing season will vary greatly from region to region and even farm to farm.

Coffee cherries go through four stages in the course of their development. In the first stage, the cherries form into small, hard balls. From around week 8 to week 26, they go through stage two, gradually growing and softening though they stay green. In stage three, which lasts from around week 26 to week 32, the cherry ripens, turning first yellow and then red. Following this point, the cherry is considered over-ripe; it darkens to a purplish color before starting to wither and turn black.

In most situations, the goal is to pick the cherries when they're in the third stage of development, without picking any cherries that are still in the first and second stage. Too many under-ripe cherries in the mix can lead to the final cup having a harsh or unpleasant vegetal taste. All the cherries on a tree don't necessarily grow and ripen at exactly the same time. Traditionally, the picking was done by hand, bringing the element of human judgment to the process. The pickers would work through the field on a rotating schedule once the harvest period started, leaving unripe cherries on the tree and coming back for them on the next pass.

The only problem with this is that hand-picking is labor intensive. Using a mechanical picker is a lot faster, especially on a large commercial farm. The problem with mechanical pickers is how to get them to discern between cherries that are ripe and those that aren't. Many farms that use mechanical pickers will also utilize a wet processing method. When the cherries are immersed in tanks of water, the unripe ones will float to the top, allowing them to be skimmed off and maintaining the quality of the beans. Farms can also carefully control the growing conditions of the beans through the use of irrigation and shade to help more of the cherries mature at the same time, leaving fewer that need to be sorted out in processing.

Hand-plucking is still the method of choice on smaller farms throughout the coffee-growing world. It is necessary on many farms at higher elevations, where the slope of the growing land prohibits the use of large machinery. As farming technology progresses, some farms are using a mix of these two methods, equipping their human pickers with specialized hand-held harvesting devices. This allows the harvesters to work more quickly while still applying their human judgment to harvest only the ripest cherries.

Some farms also use a late harvest method. This involves leaving the first cherries to ripen on the branch until they've reached the fourth stage, lessening the chances there will still be under-ripe cherries on the branch around them. This gives the finished cup a distinctive, sweet flavor. It's best used in hot, dry growing regions, where the over-ripe cherries can start to dehydrate while they're still on the tree.

Post-process sorting

Sorting the cherries before they go through any processing and drying can help to weed out immature fruits, but there are some variations in size and quality that can't be seen until the beans are freed from their cherries. Regardless of which processing and drying methods the beans go through, they will typically undergo a round of sorting following the drying.

There are two main goals at play in this final sorting: to sift out defective or low-quality beans and to separate the beans by size. Beans that are different sizes will roast at different times and temperatures; too much variation in bean size will lead to inconsistencies in the roasting stage that can diminish the quality of the final cup. As with the harvesting, sorting can be done either mechanically or by hand, with the mechanical method being quicker and hand-sorting giving more precision.

Farms may also look for specific qualities in the beans as they're sorting them. An example of this would be peaberry varieties. Most coffee beans are split into two halves, but in peaberries, these are fused into a single, round seed. Peaberries are denser than normal beans and if left in with the rest can often burn during roasting, giving the entire batch a carbon flavor if there's too many of them. When they're sorted out on their own, though, many people enjoy peaberry varieties because of the intensity of the flavor.

Processing

As you've seen in the previous chapters, the coffee cherry is intact when it comes off the tree. The seed within has to be extracted from the flesh of the cherry and then dried before it's ready to shop to roasters. The way this is accomplished will put its own stamp on the taste of the finished cup. Beans from the very same crop can taste drastically different depending on whether they're allowed to ferment within the cherry or dried first in the sun. In some locations, coffee farmers are limited by available equipment or the traditions of the region in what processing method they use; in others, farmers have more freedom to choose the processing method they think will bring out the best from the bean.

One factor that limits the kind of processing a farm can undergo is the fact that coffee cherries have to be processed immediately after they're picked. Coffee cherries left to sit for too long will begin to ferment; left to sit long enough, and this can ruin the taste of the coffee beans. Most coffee is not processed on the farm where it is grown but is instead sent to a coffee mill, essentially a plant that often processes the coffee grown on several farms in the area. Because the process has to start quickly, that farmers have to use whatever processing methods are available within quick transportation of their farm, often limiting them to whatever method is most popular in their region.

Controlled fermentation is allowed to happen in some processing methods. Fermentation refers to the naturally-occurring biochemical processes that happen when microorganisms grow in the flesh of the picked coffee cherry. During fermentation, the temperature of the bio-matter increases and the pH level drops as bacteria, yeast, and fungus break down the carbohydrates and other materials in the cherry. The trick is making sure the right bacteria are present in the mix so that the fleshy tissue of the fruit is broken down without any mold or harmful bacteria developing on or in the bean.

There are three different processing methods: dry, washed, or semi-washed. Within each of these categories, there are subtle variations in how individual mills prepare their beans. The drying stage can be done in machines, naturally under the sun, or some combination of the two. Each of these decisions will have an impact on the potential flavor of the unprocessed bean.

Dry processed

Also known as "natural processed," in the dry processing method fermentation is prevented entirely through the application of heat. The fruit is dried with the coffee beans still inside, typically by spreading them out in the sun, though it may also be done mechanically. This processing method is most prevalent in areas with a hot, dry climate, where the level of moisture in the beans can be predictably controlled; it is especially common in coffees from north and central Africa, and some of the best-known Ethiopian beans are naturally processed.

The fruit is removed from the beans once they are dried. The beans will still go through another round of drying after the removal of the fruit to make sure the moisture content is low enough for shipping. Naturally processed beans have a sweet, complex flavor with a smooth texture and a heavy body. This method is the best way to preserve the natural flavors of the bean, as the fewest chemical changes will take place inside the bean as a result.

Wash processed

In the washed processing method (also called "wet processed") the fruit and beans are allowed to ferment naturally for 12-24 hours before the fruit is removed from the seed. Typically this involves piling the cherries into large tanks, with or without the addition of water. How much water and how long the cherries are allowed to sit are up to the particular mill and farm; making changes in these areas can allow more control over the development of flavors in the bean. This process loosens the hold of the fruit's flesh on the seed; when the fermentation process is finished, the beans are sent through a machine that removes the fruit from the bean, a process known as pulping, after which the beans are dried.

Underwater fermentation brings out the acidity and aroma of the bean and tends to result in a smoother overall cup. If the process isn't done correctly, however—because of variations in the temperature of the beans or improper aeration, for example—it can give the beans an off flavor that's overly earthy, has a musty or moldy taste, or has a faint medicinal edge.

Semi-wash processed

The final processing method is a mix of the wet and dry methods and is known alternately as pulped natural or honey processed. Semi-washed processing is the newest method. It was first developed in the early 1950s but not widespread as a practice until the 1980s. The mucilage (the fleshy part of the fruit) is either only partially removed or not removed at all, allowing for some fermentation before the beans are dried. This gives it some of the sweeter, smoother notes of a washed coffee but without as much acidity.

Though honey processed is sometimes simply an alternate name for semi-washed, it can also refer to an alternate method where the coffee beans are allowed to dry naturally but in a humid environment. Particularly common in Central America, honey coffee has a high aroma and a rich body. There are three different types of honey processing: yellow honey, red honey, and black honey. In yellow honey, the mucilage is partially removed mechanically before the coffee is sun-dried, a process that takes around 8-10 days. In red honey, 50-75% of the mucilage is retained through the drying process, which lengthens the process to 12-15 days. In black honey, all of the mucilage is left on the bean throughout the drying, a process that can last up to 30 days. Black honey is the most difficult of these processes, requiring frequent mixing and rotation to prevent too much fermentation. In general, the more of the fruit is left on the bean for drying, the sweeter the resultant coffee will be.

Intestinal fermentation

Whereas the processing methods above is all done by humans, from the picking to the drying, intestinal fermentation takes a different approach. There are several animal species that eat coffee cherries as a regular part of their diet, including some species of bird, elephant, and big cats. When they eat the cherries, the seeds pass through the digestive tract, during which process the fruit is entirely cleaned off of the seed. In intestinal fermentation, the scat of these animals is collected and the beans sifted out then washed before finally being dried like normal beans. Intestinal fermentation tends to give the coffee a richer body and a stronger, lingering aftertaste with less acidity overall.

The best-known of the intestinally fermented coffees is called Luwak and is collected from the scat of the Indonesian civet cat. In some cases this is collected from wild cats; in others, they're kept on a farm. Though it first appeared in the late 1940s, it is only in recent years that it's made its way to the western coffee market. The civet cat tends to eat only the sweetest and ripest of the coffee cherries, which also ensures that the beans collected in this way are the best the tree has to offer. It takes around 12-24 hours for the seeds to pass through the animal, a similar time-span to tank fermentation methods.

Drying methods

Once the beans have been removed from the fruit, they must then be allowed to dry until the remaining moisture in the bean is between 10% and 12% to prevent mold or fungus from growing during the storage and transportation of the coffee. This drying can be accomplished one of two ways, either mechanically in large dryers or by laying out in the sun. Machine drying is a quicker process, taking only a few hours as opposed to the multiple days it takes to sun dry, but sun drying allows more of the fruity notes and natural sweetness of the bean to develop, resulting in a generally more enjoyable cup.

If the drying is not done correctly it can have a significant impact on the taste of the coffee. Drying too quickly can give the coffee a flat taste; drying too slowly, or not enough, and it can pick up a moldy or fermented flavor. Ventilation and air circulation are key in the drying process, especially for sun drying. Proper rotation of the beans ensures they all dry at the same rate and no damp spots are left once the process is over. If this process is done correctly, unroasted coffee beans can last for six months to a year when stored in a cool, dry environment.

No Harvest? No Problem!

One of the reasons that coffee flavor is so complex is that it is influenced by such a wide variety of factors. While you may not be able to start your own coffee farm in your backyard, understanding what makes coffee taste the way it does can help you make more informed decisions when you're trying out a new roast or formulating blends.

New cultivars are being developed all the time that bring unique flavors, textures, and aromas to the coffee world. If you want to stay up on the current trends, check out the Cup of Excellence winners from the past year, or check out the Specialty Coffee Association of America's website; they often have information and resources posted there that can be of great interest to coffee professionals and hobbyists alike.

Knowing what climate conditions can impact your coffee's flavor is also helpful when you're shopping for beans. If you know it's been a warmer year than usual in Mexico, you can expect the coffee from that harvest to taste a bit different than it normally does. Whether or not you want to get into things to that level of depth is up to you, of course. At the very least, understanding the harvest process will give you a deeper appreciation for how much work goes into your daily cup.

Win a free

kindle
OASIS

Let us know what you thought of this book to enter the
sweepstake at:

http://booksfor.review/harvesting

Want to

supercharge

your coffee knowledge?

Turn this page...

Brewing&Grinding Coffee

How to Make Good Coffee at Home

JESSICA SIMMS

Tasting Coffee

Coffee Cupping Techniques to Unleash the Bean!

JESSICA SIMMS

The
I know coffee
series

Steaming Milk

Want that Perfect Latte or Cappuccino?

JESSICA SIMMS

Making Crema

The Art and Science of the Perfect Espresso Shot

JESSICA SIMMS